GOAT FARMING FOR BEGINNERS

EASY GUIDE

Uncover Expert Insights On Breeds Selection,
Optimal Housing, Nutritional Mastery, Health
Management Tactics, And Proven Marketing
Strategies For Success

WAYLEN BANNARD

INTRODUCTION

In recent years, goat husbandry, a conventional agricultural method, has garnered considerable interest owing to its prospective profitability and sustainability. The purpose of this introductory section is to offer an overview of goat farming, including its historical context, the significant role it plays in agriculture, and the precise goals that this all-encompassing manual endeavor to accomplish.

Background (1)

Comprehending the historical backdrop of goat husbandry is critical to fully grasp its transformation into a contemporary and profitable enterprise. Since their domestication thousands of years ago, goats have provided flesh, milk, and fiber. Across various historical periods and in present-day societies, goats have consistently served as vital providers of sustenance and economic stability. Through an examination of the origins of goat husbandry, one can discern the customary methodologies that established the groundwork for the subsequent expansion and metamorphosis of the sector.

2 Importance of Goat Farming

Goat farming is of significant significance within the agricultural domain due to a multitude of factors. Goats, to begin with, are multipurpose animals that make contributions to numerous industries, such as the production of meat, dairy, and fiber. Additionally, goats possess a reputation for clinging to various terrains and climates, which renders them viable for agricultural purposes across a broad spectrum of surroundings.

Further, goat husbandry is frequently regarded as a sustainable method because goats can flourish in marginal areas where other livestock might have difficulty. This segment will explore the diverse and substantial implications of goat husbandry, with a particular focus on its contributions to environmental sustainability, rural livelihoods, and food security.

3 Objectives of the Book

Clear objectives are crucial for achieving success in any undertaking. The primary objective of this book is to provide an all-encompassing manual of methods for individuals who are contemplating goat husbandry or are interested in improving their current operations.

The stated objectives include imparting practical knowledge about the administration of goats, breeding methodologies, healthcare approaches, and marketing tactics. Through an examination of the distinct requirements and obstacles encountered by goat farmers, this literary work endeavors to furnish readers with the understanding and competencies essential for operating a lucrative and environmentally conscious goat farming enterprise. This segment will delineate the primary objectives and anticipations that readers may encounter as they progress through the following chapters, thereby cultivating a feeling of intention and guidance.

CHAPTER ONE

INTRODUCTORY SECTION REGARDING GOAT FARMING

1. Choosing the Appropriate Breed

It is crucial that you choose the proper breed to ensure the success of your goat husbandry enterprise. Various varieties possess unique attributes that may exert an impact on their capacity to adjust, rate of development, and overall output. Perform extensive investigation on breeds that are compatible with your particular agricultural objectives and the local climate. Considerations include the breed's disease resistance, reproductive capabilities, and potential for producing meat or milk. Consult veterinary experts, interact with seasoned farmers, and participate in community agricultural gatherings to acquire knowledge regarding the most appropriate breed for your particular geographical area.

2 Establishing the Infrastructure of the Farm

Establishing an optimal and thoughtfully planned agricultural infrastructure is critical for ensuring the general welfare and efficiency of your goat herd. The design of your farm should encompass the precise placement of feeding areas, shelters, and grazing areas. Assure that the infrastructure facilitates effective drainage and ventilation to avert health complications. Establish robust and impervious fencing to protect the goats from potential external hazards. Sufficient housing and hygienic, orderly dining areas are elements that promote the comfort and well-being of the goats. An impeccably planned agricultural configuration not only optimizes productivity but also serves as a pivotal element in germ control.

Three Supplies and Equipment

It is critical to furnish your goat farm with the proper equipment and supplies to ensure effective management and upkeep. Invest in water troughs and feeders, among other appropriate feeding apparatus, to guarantee that the goats are adequately nourished. Grooming equipment, hoof trimmers, and medical supplies are considered indispensable instruments for regular maintenance. Efficient handling facilities, such as pens and chutes, simplify arduous duties like health checks and vaccinations. Inspect and maintain your equipment on a routine basis to ensure its durability and dependability. Furthermore, it is imperative to cultivate dependable suppliers for feed, medications, and other critical items to ensure a steady and high-quality supply chain.

4 Legal Factors to Consider

It is essential to comprehend and abide by the local laws and regulations governing livestock farming before beginning goat farming. Acquire the requisite permits and licenses to lawfully operate a goat farm. Educate yourself on the environmental guidelines, zoning regulations, and animal welfare standards that may affect your agricultural practices. Legal counsel or local agricultural extension services should be consulted to ensure complete adherence to all pertinent legislation. By proactively confronting legal considerations, one can proactively avert potential complications and lay the groundwork for a goat husbandry enterprise that is both profitable and sustainable. Maintain a consistent awareness of any modifications or additions to regulations to adjust your practices accordingly.

CHAPTER TWO

GOAT NUTRITION AND FEEDING

1. Comprehending Dietary Requirements

The nutritional status of goats is critical for assuring the productivity and overall health of a goat farming operation. A comprehensive comprehension of the nutritional needs of goats is essential to design an economically viable and harmonious feeding regimen.

Essential constituents of a goat's dietary regimen comprise carbohydrates, proteins, lipids, minerals, and vitamins. Carbohydrates procured from grains and forages serve as a source of energy, whereas proteins, which are derived from grains and legumes, facilitate muscle growth. Fats, which are obtained from cereals and oils, aid in the absorption of vitamins and energy. A thorough comprehension of the distinct nutritional requirements throughout various life phases, including gestation, lactation, and growth, is essential to maximize the health and productivity of goats.

<u>Two Varieties of Forage and Feed</u>

Selecting the proper feed and forage is a crucial factor in determining the profitability of goat husbandry. Hay, cereals, and legumes are among the diverse feeds that comprise a goat's diet. Forage of superior quality, including alfalfa and clover, provides vital nutrients. Grooms also derive nutritional value from a variety of cereals, including corn, barley, and oats. Consistency in the proportion of roughage to concentrate is critical for the preservation of digestive well-being. Implementing rotational grazing systems can guarantee that animals have access to a variety of forage options.

Comprehending the nutritional makeup of various fertilizers and forages is critical in devising a balanced diet that adequately satisfies the particular requirements of goats.

Three Feeding Methods to Promote Optimal Growth

It is critical to implement efficient nutrition methods to facilitate optimal growth in goats. Ensuring proper nutrition management entails maintaining a balanced and consistent diet on an annual basis. Feeding goats by their age, weight, and stage of production is essential. The utilization of nutrients is enhanced when the daily ration is divided into multiple feedings.

Ensuring access to potable water is critical for optimal digestion and general well-being.

To prevent malnutrition and obesity, adjusting feeding strategies by body condition scores is beneficial. Further, by comprehending the temporal fluctuations in the quality of forage and modifying the goats' diet accordingly, one can guarantee that they acquire the essential nutrients required for optimal development and procreation.

Minerals and supplements are essential for addressing nutrient deficiencies and improving the general health of goats. Goats necessitate particular minerals for processes including reproduction, bone development, and immune function, including calcium, phosphorus, and selenium. Deficit identification via analysis of soil and forage facilitates the determination of the most suitable supplementation. Mineral blocks or loose mineral mixtures guarantee that goats have unrestricted access to vital nutrients.

Vitamins, including A, D, and E, play an essential role in numerous physiological processes and ought to be integrated into the diet. Ensuring the appropriate provision of supplements and minerals is crucial for mitigating health complications and bolstering the financial viability of the goat husbandry enterprise.

CHAPTER THREE

MANAGEMENT OF BREEDING

Section I: Breeding Stock Selection

Selection of breeding stock is an essential component of prosperous goat husbandry.

The enterprise's profitability is substantially reliant on the caliber of the reproductive animals. When selecting breeding stock, prioritize characteristics including high milk production, disease resistance, and excellent conformation. Select animals from reputable breeders or trustworthy sources that have established a successful history. Consistent genetic screening and health examinations can assist in the detection of possible reproductive complications and guarantee a strong herd genetic base for your goats.

2. The Estrus Cycle and the Health of Reproduction

Comprehending the estrus cycle and upholding reproductive health in goats are critical factors in ensuring effective reproduction management. Because it can differ, acquaint yourself with the typical estrus cycle of your goat breeds. Consistent observation of the does for indications of heat, including heightened vocalization and mounting behavior, facilitates the determination of the most advantageous periods for reproduction. Establish a proactive healthcare initiative aimed at resolving prevalent reproductive concerns and safeguarding the reproductive health of does. Vaccinations, proper nutrition, and parasite control are all substantial contributors to reproductive health.

Three Mating Methods

The implementation of efficient mating techniques is vital in guaranteeing favorable reproduction results. Age, weight, and health status must all be taken into account when determining the most suitable mating pairs.

By introducing specific males to does at predetermined intervals, controlled mating can facilitate the management of breeding schedules and guarantee a more consistent birthing season. Additionally, precise genetic enhancement may be accomplished through the use of artificial insemination (AI). Maintaining accurate records of mating dates and results enables the formulation of strategic breeding plans for future seasons.

4. Pregnancy and the Management of Kidding

After effective reproduction, careful management of pregnancy and kidding becomes crucial. Administer adequate nutrition to expectant women, ensuring they have sufficient quantities of vital minerals and vitamins. Consistent veterinary examinations facilitate the monitoring of the health of the does and enable timely intervention in any potential complications. As the due date approaches, ensure that the delivery environment is comfortable and sanitary.

Throughout the birthing process, maintain vigilance for any indications of complications and ensure you have the required supplies on hand. Emphasize appropriate child care after the birth of a child, including colostrum consumption and preventative healthcare measures to enhance the health of the herd as a whole.

In essence, effective breeding management in the goat farming industry necessitates meticulous attention to reproductive health, mating techniques, breeding stock selection, pregnancy and offspring management, and reproductive health. The diligent implementation of these practices will ultimately result in a robust and thriving goat herd, thereby optimizing the financial gains of the enterprise.

CHAPTER FOUR
HEALTH AND DISEASE MANAGEMENT

1. Frequent Goat Illnesses

Goats, similar to other livestock, are vulnerable to a wide range of maladies that have the potential to greatly affect their well-being and efficiency. It is imperative to comprehend prevalent goat diseases to regulate their health effectively on a goat farm. "Caprine Arthritis Encephalitis (CAE)," a viral infection that impacts both the joints and nervous system, is a commonly observed condition. Among the symptoms are neurological disorders and disability. "Caseous Lymphadenitis (CL)," a bacterial infection that results in lymph node abscesses, is an additional cause for concern. Maintain proper sanitation in the barn and quarantine new goats before introducing them to the herd to prevent the spread of disease. Consistent health examinations and timely isolation of ill goats are critical protocols for the efficient management of prevalent goat diseases.

2. Preventive Actions

Disease prevention in a goat husbandry operation is dependent upon proactive health management. It is critical to implement biosecurity measures to prevent the introduction and proliferation of pathogens. In goats, proper nutrition significantly contributes to the enhancement of their immune systems. Supply an adequately balanced diet that includes essential vitamins and minerals. Ensuring adequate sanitation practices, such as routinely cleaning pens and apparatus, is essential for preserving a healthy living environment. Managing stressors such as congestion and sudden dietary changes is also crucial for preventing the development of diseases. Through the implementation of these preventative measures, goat producers can establish a formidable barrier against potential health complications.

3. Animal Care

Consistent veterinary attention is vital to the success of goat husbandry. Promoting a positive professional rapport with a certified veterinary practitioner guarantees timely identification and management of any medical concerns.

Consistent health examinations, which encompass the assessment of body weight, condition, and fecal counts, serve to detect potential issues before their exacerbation. It is essential to address the dental health of a goat, as dental problems can impair its ability to eat correctly. An additional service provided by a veterinarian is the development of a health management plan that is specific to the goat herd's requirements. Prompt veterinary interventions, in conjunction with preventative measures, make a substantial contribution to the goats' general welfare and the farm's financial viability.

4. Vaccination Schedule

The implementation of a systematic vaccination schedule is critical in safeguarding goats against avoidable illnesses. Routine administration of core vaccines, including those targeting Clostridium perfringens and Pasteurella, is recommended. Vaccination against tetanus toxoid is of the utmost importance, particularly for goats undergoing dehorning or other surgical procedures. Adapt the vaccination regimen to the goat herd's particular requirements and the prevalence of diseases in the area. Verify that the vaccination schedule is consistent with the objectives of the farm and regional health considerations by consulting a veterinarian. Consistently evaluate and revise the vaccination protocol to proactively mitigate emergent pathogen risks and preserve peak herd immunity, thereby making a substantial contribution to the overall financial viability of the goat farming enterprise.

CHAPTER FIVE

GOAT HOUSING AND MANAGEMENT

1. Constructing Appropriate Shelters:

The construction of an appropriate shelter is critical for promoting the health and efficiency of goats within an agricultural operation. Goats must be protected from inclement weather, including extreme cold, precipitation, and heat, by the shelter's strategic placement. Sufficient ventilation is critical for preserving air quality and averting respiratory complications. Furthermore, the construction materials must exhibit durability, cleanliness, and resistance to external elements. Sufficient area within the shelter is essential to adequately house the herd. The implementation of a meticulously planned shelter design significantly improves the overall health and alleviates tension among goats, consequently leading to increased productivity and profitability.

2. Cleaning and Bedding Procedures:

The implementation of proper bedding and cleansing protocols is crucial for ensuring that goats inhabit a hygienic and disease-free environment. Selected bedding material, such as wood sawdust or straw, not only offers solace to the goats but also facilitates the process of waste absorption. Consistent sanitation and elimination of contaminated linens serve to avert the buildup of pathogens and parasites. Proper waste management system implementation is critical for preserving sanitation. Goat producers can prevent the spread of diseases and safeguard the herd's overall well-being and efficiency by strictly adhering to cleaning protocols. Maintaining a high standard of cleanliness not only ensures the welfare of the goats but also positively impacts the financial viability of the agricultural enterprise.

3. Space Management

Effective space management is critical for safeguarding the physical welfare of goats and preventing overload. Sufficient space is essential for the movement, resting, and nourishment of each goat. The consequences of overcrowding include increased tension, the transmission of diseases, and decreased productivity.

The configuration of the goat housing area should be meticulously planned by farmers, taking into account various factors including the breed, herd size, and the distinct requirements of individual goats. The adoption of effective space management strategies not only improves the general well-being of the goats but also has a beneficial impact on their growth, reproduction, and production of milk or flesh, thereby ultimately enhancing the financial viability of the goat farming enterprise.

4. Environmental Points to Consider:

A sustainable and environmentally friendly system for goat husbandry necessitates the incorporation of numerous environmental factors. Composting is one waste management strategy that facilitates the recycling of organic matter and reduces environmental impact.

Additionally, the use of renewable energy sources for heating and ventilation systems within the goat shelters should be considered. Planting vegetation around the farm can also promote a more natural habitat for the goats and contribute to environmental conservation by providing shelter. Farmers must maintain an equilibrium between the welfare of the livestock and the conservation of the environment. Goat producers can contribute to ecological sustainability and enhance the marketability and reputation of their products in an ever more environmentally aware consumer base by incorporating environmentally conscious practices.

CHAPTER SIX
MARKETING AND SALES STRATEGIES

1. Target Market Identification: Comprehending the Customer Base

Target market identification is a crucial component in the formulation of effective marketing and sales strategies for the goat husbandry industry. To commence, undertake an extensive market analysis to identify prospective clientele who possess a need for goat-derived merchandise. Consider psychographics, demographics, and geographic location, among other factors. Rural communities might be more interested in goat flesh, whereas specialty goat cheeses might attract a greater number of urban consumers. Customizing products to meet the distinct requirements of individual customers will not only augment customer contentment but also optimize marketing endeavors, thereby maximizing financial gains.

2. Establishing a Distinctive Identity for Your Goat Farm

The significance of branding in distinguishing your goat farming enterprise from rivals cannot be overstated. Construct an alluring brand that accurately represents the farm's values, the superior quality of its products, and the distinctive characteristics of its operation. For your goat products, you should consider devising appealing packaging and developing a memorable logo. Establishing a solid brand identity cultivates client allegiance and confidence, thereby establishing your goat farm as a reputable and favored option within the marketplace. By relating the narrative, implementing sustainable practices, and establishing a personal connection with consumers, you can increase the perceived value of your products.

3. Sales Channels: Expanding the Reach to a More Diverse Audience

Investigate a variety of sales channels to increase sales and expand your market reach. One potential avenue for direct consumer sales is via online platforms, farmers' markets, or on-farm stores. Investigate potential alliances with regional supermarkets, eateries, and specialty food retailers. The implementation of diverse sales channels not only guarantees a consistent flow of revenue but also affords enhanced adaptability to shifts in the market. By capitalizing on the potential of digital marketing and online sales, utilize social media platforms and e-commerce to establish connections with a broader audience and exhibit your goat farming products to prospective clients located in different geographic regions.

4. Pricing Strategies: Balancing Demand and Profitability in the Market

In goat husbandry, pricing strategies are critical for ensuring profitability and maintaining a competitive edge. Perform an exhaustive examination of production expenses, market demand, and rival pricing to establish prices for your goat products that are both competitive and profitable. One potential strategy to accommodate a wide range of consumer preferences is to incorporate tiered pricing according to product variants or packaging sizes. To incentivize sales, implement promotional pricing during designated seasons or events. Consistently evaluate and modify your pricing strategy to conform to market dynamics and guarantee the long-term financial viability of your goat farming enterprise.

CHAPTER SEVEN

FINANCIAL MANAGEMENT

1. Cost Evaluation

Introduction: Cost analysis is a critical component in determining the profitability of goat husbandry. Gaining knowledge of the diverse expenditures associated with goat farming empowers producers to make well-informed decisions and maximize the utilization of resources. Expenses may include sustenance, labor, veterinary care, infrastructure, and equipment, among others. A comprehensive examination of costs facilitates the identification of opportunities to increase efficiency, thereby bolstering the overall financial performance.

Feed expenditures comprise a substantial proportion of the total costs associated with goat husbandry. Managing expenses requires that the nutritional requirements of goats be evaluated and that cost-effective feed options be investigated. By comparing the cost and quality of various feed sources, producers can achieve a balance between feeding the goats optimally and maintaining cost control.

Expenses Associated with Labor and Infrastructure: Goat husbandry necessitates labor for infrastructure maintenance, breeding, and daily care, including the upkeep of shelters and fences. An exhaustive cost analysis takes into account labor costs, building materials, and recurring maintenance expenditures. Farmers can efficiently manage their operations and minimize superfluous expenditures by acquiring knowledge of these costs.

Veterinary care is an essential component of goat husbandry, and the financial implications associated with it are substantial. By conducting a cost analysis of vaccinations, medications, and routine health check-ups, producers can develop a proactive healthcare strategy. The timely identification and mitigation of potential health hazards are crucial for safeguarding the goat herd against financial losses and ensuring its overall welfare.

2. Allocation of Funds for Goat Farming

Introduction: Budgeting serves as a proactive instrument of financial planning that facilitates the efficient allocation of resources for goat producers. Estimating income and projecting expenses comprise the process of developing a comprehensive budget, which serves as a road map to financial success. An effectively crafted budget functions as a crucial instrument for facilitating informed decision-making and adjusting to dynamic market circumstances.

Income Projection: The process of income estimation entails the assessment of prospective sources of revenue, including the trade of goats, milk, or other by-products. When endeavoring to forecast income, farmers must take into account market prices, fluctuations in demand, and the possibility of livestock growth. Adopting a proactive stance aids in establishing attainable financial objectives and facilitating well-informed investment choices.

Expense Projection: A crucial component of budgeting is the anticipation of expenses. Farmers can develop a comprehensive forecast of forthcoming expenditures by integrating cost analysis data with market trends and inflation rates. This foresight enables proactive cost management and resource allocation, thereby securing the financial stability of the goat husbandry enterprise.

Contingency Planning: An all-encompassing budget incorporates provisions to address unanticipated events and urgent situations. This entails allocating funds in anticipation of unforeseen veterinary costs, market volatility, or natural calamities. By incorporating contingency planning into their budget, producers can enhance their resilience and flexibility, thereby gaining the confidence to navigate uncertainties.

3 Record Maintenance

Effective recordkeeping is fundamental to the success of goat husbandry, as stated in the introduction. Farm management entails the methodical recording of diverse facets of the farm's activities, which empowers proprietors to monitor progress, render decisions based on data, and adhere to regulatory obligations. Comprehensive documentation offers valuable insights regarding the financial well-being, efficiency, and potential areas for enhancement of the property.

Herd management records are of utmost importance as they facilitate the effort to maintain accurate information regarding the birthdates, health histories, and reproductive activities of every goat. This data assists in monitoring the herd's overall well-being and efficiency, enabling prompt interventions and optimizing breeding initiatives to achieve desired characteristics.

The maintenance of precise financial records is critical to monitor and assess revenue, expenditures, and overall profitability. This encompasses invoices, receipts, and statements about various expenses such as feed, labor, and veterinary care. Financial records are essential for tax compliance and obtaining financial assistance, as they provide a transparent picture of the farm's financial health.

The documentation of production metrics, including but not limited to milk yield, reproduction rates, and development rates, provides significant insights into the overall productivity of the farm. Through the analysis of these metrics over some time, producers can discern patterns, assess the efficacy of breeding initiatives, and arrive at well-informed decisions that optimize overall production efficiency.

Optimising Profits (4)

Introduction: To optimize profits in goat husbandry, a strategic approach is necessary, encompassing resource optimization, productivity enhancement, and the exploration of diversified income streams. Goat producers can maximize the benefits of their enterprise by emphasizing critical domains including marketing, herd management, and value-added products.

A comprehensive comprehension of market trends, consumer preferences, and demand dynamics is imperative to optimize financial gains. To succeed, farmers must discern specialized markets, devise efficient promotional tactics, and adjust to evolving consumer inclinations. Furthermore, by investigating potential avenues for direct sales and establishing partnerships with regional enterprises, one can potentially enhance their financial gains.

The optimization of breeding programs is a critical component in the pursuit of profit maximization. Prioritize the propagation of goats with desirable characteristics that correspond to market demands. This process entails meticulous breeding pair selection, diligent genetic diversity monitoring, and the implementation of breeding strategies aimed at augmenting the herd's overall quality.

To optimize profitability and reduce risks, farmers should consider the implementation of diversified income sources. Potential offerings could consist of goat flesh, milk, and hides for sale, as well as agritourism experiences. By utilizing a variety of revenue streams, producers can establish a business model that is more robust and resistant to changes in the market.

prosperous goat husbandry necessitates an amalgamation of astute financial administration, strategic forethought, and ongoing adjustment to fluctuations in the market. Goat producers can establish a profitable and sustainable enterprise over an extended period through the implementation of optimal profit maximization strategies, comprehensive cost analysis, and efficient budgeting.

CHAPTER EIGHT
SUSTAINABLE AND ORGANIC PRACTICES

1. Organic Goat Production

Sustainable goat husbandry is characterized by the absence of synthetic chemicals, hormones, and genetically modified organisms during goat production. This approach emphasizes the welfare of both animals and the environment among cultivators. Maintaining a natural and toxin-free environment, ensuring appropriate animal welfare, and supplying organic feed are the guiding principles. Farmers not only ensure the production of superior ruminant products but also make a positive environmental impact by adhering to organic principles, which safeguard soil fertility and biodiversity. The objective of this farming approach is to establish a symbiotic relationship between industrial profitability and ecological accountability.

2. Sustainable Practices

Agricultural practices that are environmentally friendly aim to reduce the ecological consequences of agricultural operations. This includes a variety of approaches, including the conservation of energy and water, the reduction of pollution, and the preservation of soil. Agroforestry, crop rotation, and integrated pest management are sustainable agricultural practices that reduce the demand for chemical inputs and aid in biodiversity conservation. Implementing systems that foster climate change resilience and ensuring the responsible use of resources, including land and water, are all components of environmentally sustainable goat husbandry. Implementing these practices not only serves to protect the environment but also improves the farm's economic sustainability through the reduction of input expenses and the enhancement of long-term viability.

3. Certification Methods

In goat farming, certification processes are vital for establishing the credibility of organic and environmentally favorable practices. Compliance with particular criteria established by certifying agencies or regulatory bodies is required to obtain certification. Certification guarantees adherence to organic principles in the context of goat husbandry, encompassing the utilization of organic feed, the provision of humane treatment for animals, and a steadfast commitment to environmental conservation. Eco-friendly certifications may incorporate a wider range of criteria, assessing the comprehensive environmental impact of the farming operation. By obtaining these certifications, farmers gain access to premium markets that prioritize sustainability and earn the trust of consumers. On the contrary, the certification procedure necessitates thorough documentation, strict adherence to guidelines, and regular audits to guarantee continuous compliance with organic and sustainable practices.

certification processes, eco-friendly practices, and organic goat farming are interdependent components that significantly impact the profitability and long-term viability of contemporary agricultural enterprises. Farmers who emphasize ethical and environmentally conscious practices not only ensure the production of superior-quality goods but also correspond with the preferences of consumers who favor sustainable and organic alternatives. The integration of these methodologies not only facilitates financial prosperity but also significantly contributes to the development of a sustainable and environmentally conscientious agricultural sector.

CHAPTER NINE
CHALLENGES AND SOLUTIONS

1. Frequent Obstacles in Goat Farming

Despite its considerable financial gain, goat husbandry is not devoid of its share of difficulties. Disease control is a significant obstacle encountered by goat producers. Goats are vulnerable to a range of diseases, including parasitic infestations and respiratory infections, both of which have the potential to substantially compromise the herd's overall health. Moreover, maintaining adequate sustenance for the goats is an ongoing obstacle, given that their dietary requirements may differ according to variables such as breed, age, and production phase. Additionally, substandard lodging facilities and infrastructure may contribute to stress and hygiene issues among the goats, thereby exacerbating health complications.

Farmers may face financial obstacles due to market volatility and the unpredictability of goat product demand.

2. Problem-Solving Methods

To surmount the obstacles encountered in goat husbandry, proactive measures and strategic planning are required. The prevention of diseases can be achieved by implementing routine veterinary examinations, immunizations, and stringent biosecurity measures. Animals must be fed a nutritionally adequate and well-balanced diet; therefore, producers should contemplate seeking guidance from animal nutritionists to develop appropriate feeding strategies.

Farmers can address housing and infrastructure challenges by allocating resources towards the construction of suitable shelters that feature adequate ventilation, drainage, and sanitation maintenance. Furthermore, implementing a strategy of product diversification—including milk and other by-products in addition to meat— can serve to alleviate the effects of market volatility and guarantee a more consistent stream of revenue.

Engaging in collaborative efforts with fellow farmers and industry experts to exchange knowledge can yield significant insights about efficacious problem-solving methodologies.

Future Developments In Goat Farming

Numerous emerging trends have the potential to significantly transform the goat husbandry industry in the coming years. The increasing recognition of sustainable and organic agricultural methods can be attributed to the growing consumer preference for environmentally conscious and ethically produced goods. Increasing numbers of goats are being monitored through the use of sensors and other monitoring devices, which contributes to the expansion of technology integration. With the assistance of data analytics and automation, precision agriculture can maximize the use of resources and improve overall farm productivity.

Furthermore, the increasing demand for unique and specialized products, including goat cheese and specialty breeds, provides producers with fresh opportunities to investigate and broaden their business endeavors. With the ongoing increase in global demand for protein sources, goat husbandry is anticipated to receive heightened attention, which will create favorable conditions for expansion and financial gain.

Finally, to overcome the obstacles associated with goat husbandry, a comprehensive strategy is necessary, which includes health management, nutrition, infrastructure, and market dynamics. By adopting proactive problem-solving strategies and remaining cognizant of forthcoming trends, goat producers can establish a solid foundation for long-term success in this dynamic industry.

CHAPTER TEN
SUCCESS STORIES AND CASE STUDIES

1. Biographies of Prominent Goat Farmers

When investigating prosperous goat producers, it is critical to examine the biographies of individuals who have established a distinct presence in the field. An exemplary case study involves Mr. John Anderson, a seasoned goat farmer whose farm has achieved unprecedented success due to his meticulous management practices.

By implementing sustainable agricultural practices, Anderson prioritized breed selection to ensure that his livestock consisted of superior goats possessing desirable characteristics.

The individual's dedication to ongoing education and embracing contemporary technologies significantly contributed to the attainment of peak productivity.

Through cultivating robust relationships with regional markets and adopting ethical business principles, Anderson established a viable framework that not only enhanced his financial gains but also played a pivotal role in the expansion of the goat farming industry as a whole.

2. Insights Gained

An analysis of the insights gained by prosperous goat producers illuminates the fundamental tenets that govern success in the sector. An essential takeaway is the significance attributed to effective livestock management.

Efficient farmers place a high value on the welfare and health of their goats, as evidenced by their adherence to stringent vaccination protocols and the employment of proficient veterinarians. Moreover, towards achieving success, effective pasture management and feed optimization strategies are essential elements. An additional crucial lesson pertains to the implementation of strategic marketing and sales techniques. Proficient goat producers discern market trends with skill, thereby guaranteeing that their products satisfy the demands of consumers. By incorporating these principles into their practices, individuals aspiring to become goat producers can strengthen their fundamental principles and confront obstacles with fortitude.

3. Motivational Tales

Within the domain of goat husbandry, there is a profusion of motivational anecdotes that illustrate the success of perseverance and ingenuity. An illustrative example is the account of Ms. Grace Mbeki, an unmarried mother who successfully converted a humble goat property into a prosperous business. Despite encountering various obstacles such as resource scarcity and societal skepticism, Mbeki demonstrated resolute determination in his pursuit.

By utilizing government assistance programs and establishing connections with agricultural professionals, she systematically grew her property. The narrative serves as a poignant illustration of how determination and passion can profoundly alter the course of challenges encountered in the goat farming industry. Motivating narratives such as Mbeki's function as beacons of hope, rousing ambitious farmers to fervently pursue their aspirations.

CONCLUSION

The inspirational narratives, success stories, and lessons learned that comprise the goat farming industry as a whole provide aspiring farmers with a wealth of knowledge and inspiration.

The profiles of prosperous goat producers serve as prime examples of the concrete outcomes that can be attained by integrating knowledge, commitment, and environmentally conscious methodologies. The insights gained emphasize the criticality of effective livestock management and strategic marketing approaches in guaranteeing the success of a goat farming enterprise.

Motivational narratives, exemplified by the experiences of Ms. Grace Mbeki, shed light on the capacity for profound change and achievement despite confronting formidable obstacles. Prospective goat farmers may find inspiration in these anecdotes to guide their endeavors, thereby contributing to the goat farming industry's long-term sustainability and success.

www.ingramcontent.com/pod-product-compliance
Lightning Source LLC
Chambersburg PA
CBHW070824290526
45795CB00002B/838